EXPLORING THE SCIENCE OF NATURE

The Nature and Science of
RAIN

Jane Burton and Kim Taylor

Gareth Stevens Publishing
MILWAUKEE

For a free color catalog describing Gareth Stevens Publishing's list of high-quality books and multimedia programs, call 1-800-542-2595 (USA) or 1-800-461-9120 (Canada). Gareth Stevens Publishing's Fax: (414) 225-0377. See our catalog, too, on the World Wide Web: http://gsinc.com

Library of Congress Cataloging-in-Publication Data

Burton, Jane.
The nature and science of rain / by Jane Burton and Kim Taylor.
p.   cm. — (Exploring the science of nature)
Includes index.
Summary: Explains how vital rain is, its chemical composition, the water cycle, how animals waterproof themselves, and ways of measuring and making rain.
ISBN 0-8368-1944-6 (lib. bdg.)
1. Rain and rainfall—Juvenile literature. [1. Rain and rainfall.] I. Taylor, Kim.
II. Title. III. Series: Burton, Jane. Exploring the science of nature.
QC924.7.B85 1997
551.5'77—dc21   97-8480

First published in North America in 1997 by
Gareth Stevens Publishing
1555 North RiverCenter Drive, Suite 201
Milwaukee, Wisconsin 53212 USA

This U.S. edition © 1997 by Gareth Stevens, Inc. Created with original © 1997 by White Cottage Children's Books. Text and photographs © 1997 by Jane Burton and Kim Taylor. The photograph on page 18 *(top)* is by Mark Taylor. Conceived, designed, and produced by White Cottage Children's Books, 29 Lancaster Park, Richmond, Surrey TW10 6AB, England. Additional end matter © 1997 by Gareth Stevens, Inc.

The rights of Jane Burton and Kim Taylor to be identified as the authors of this work have been asserted by them in accordance with the Copyright, Design and Patents Act 1988. Educational consultant, Jane Weaver; scientific adviser, Dr. Jan Taylor.

Printed in the United States of America

1 2 3 4 5 6 7 8 9 01 00 99 98 97

Contents

Words that appear in the glossary are printed in **boldface** type the first time they occur in the text.

Miraculous Rain

Rain is essential to our planet. Heavy rain can be exciting, but also dangerous. Raindrops come pelting down, the ground gets soaked, and puddles form. Swirling water fills the rivers and may burst over the banks. But after a heavy rain when the Sun shines again, leaves are fresh and green, birds sing, and flowers open. Rain is absolutely necessary to life on Earth.

Rain occurs because air and water are able to mix. It may seem unlikely that a gas (the air) and a liquid (the water) can mix. But that is exactly what happens when water evaporates into the air, and that is where rain begins to form.

Water is made up of two gases, **hydrogen** and **oxygen**. Two **atoms** of hydrogen combine with one atom of oxygen to make one **molecule** of water. The **chemical formula** for water is H_2O.

Below: A raindrop falls into a pool of water causing a spike of water to shoot up and break into smaller drops.

Opposite: Dark clouds appear along a coastline, blocking the Sun. Rain falls in gray curtains over the sea.

Wet Air

Top: The brilliant colors of harlequin bugs gleam among the moist green leaves of the **tropical** forest.

There is always water in the air. It is in the form of molecules that float around and mix with molecules of air. The water is in a gas-like form known as water **vapor**. There is always some water vapor in the air.

6

In desert areas, there is just a little. But in **rain forests**, there is an abundance. The amount of water vapor in the air is measured as a percentage. This measurement is known as the **relative humidity**. When the relative humidity is 100 percent, the air cannot carry any more water. In other words, the air is **saturated**.

Above: The green iguana gets the water it needs by eating leaves that contain water and by licking raindrops off **foliage**.

Left: Plants take water from the ground through their roots. In turn, they release water vapor into the air through their leaves. Where plants are plentiful, as in the rain forest, the air is always moist.

Fog and Clouds

Have you ever walked near a river on a misty morning or walked through a city fog? If you have, you actually walked through tiny specks of water that probably could not be seen. These tiny water droplets are much easier to see in bright light, especially when the Sun shines.

Warm air carries more water vapor than cold air. When air is cooled, its relative humidity rises. If it is cooled enough, the air may become saturated. Then, the vapor **condenses** into water droplets. When moist air near the ground rises high into the cooler regions of the **atmosphere**, clouds form. When the ground becomes cold and cools the air next to it, fog forms.

Opposite: On a damp, misty autumn morning, the air is saturated with water vapor. Fine droplets hang in the air. Every surface is wet, including the twigs and leaves of the trees. The forest is filled with the sound of a steady drip, dripping.

Below: Mist forms close to the ground when the air cannot hold any more water vapor. Some of it condenses, and every surface — even a piece of thistledown caught in the web of a spider — becomes laden with water drops.

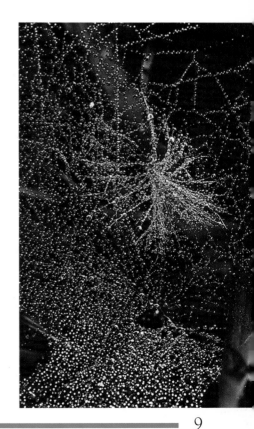

Opposite: Clouds form when wind carries moisture-laden air over mountains. The wet air is pushed up and cooled, causing the vapor to condense into droplets.

Below: The shallow water visited by these greater flamingos was warmed all day by the Sun. Now it is evening, and the incoming tide is so cold that it cools the air near it, forming a mist.

Most clouds are made up of millions of tiny water droplets **suspended** in the air. The water collects in the air in the form of vapor. Much of it comes from the surface of the sea, particularly from warm, tropical seas. This is because warm water vaporizes easier than cold water. When saltwater from the sea evaporates, the salt is left behind.

But cold seas also supply water vapor to the air, especially when whipped up by stormy winds. A cold tide, like the one off the coast of Namibia in southwestern Africa *(below)*, puts a large amount of water vapor into the air.

Falling Rain

What causes a cloud to turn into rain? The answer depends on the size of the water droplets the cloud contains. When the droplets grow beyond a certain size, they fall to the ground.

The reason droplets in a cloud sometimes grow is mostly a matter of temperature. When a cloud becomes colder, more water vapor condenses onto the droplets. The droplets get bigger and drift downward. The falling droplets gather even more water on their way down, soon becoming raindrops that tumble to Earth.

Right: Raindrops falling from a thundercloud collect moisture on their way to the ground. The farther they fall, the bigger they get and the faster they travel. The large ones hit the ground hard and quickly form puddles.

Clouds cool when the air that carries them pushes higher into the atmosphere. A good example of this is when the wind blows clouds over mountains. Mountainous areas close to the sea are some of the wettest places on Earth. Over very high mountains, the air is so cold that snow falls from the clouds.

Above: Thunderstorms often form when currents of warm, moist air rise high in the sky above ground that has been heated by the Sun.

Drips and Drops

Water droplets in clouds and small raindrops that drift gently toward Earth are perfect spheres. They are like tiny round beads. Each drop is held in its shape by what is known as **surface tension**. It is as if there were a piece of elastic keeping each drop round. All water surfaces — even the flat surfaces of lakes and puddles — seem to have this invisible elastic over them.

As falling raindrops get bigger, they hurtle through the air faster and their shapes change. The rush of air pushing against their lower edge makes them egg shaped. Eventually, surface

Below: A drop falls off a berry.

It hits the surface of a pond.

The drop makes a hollow in the surface of the pond.

Water rushes from below to fill the hollow but goes too far and forms a spike.

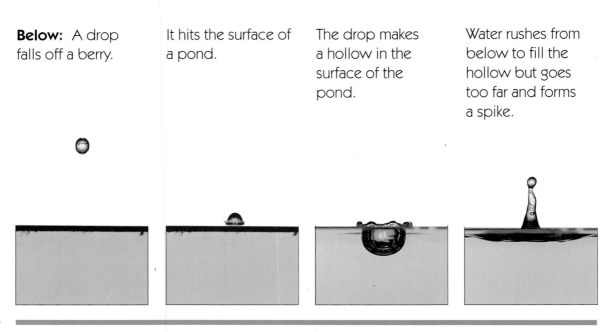

Left: Raindrops falling into a pond hit the surface so hard that little spikes of water form and jump out of the pond. A spike with a drop above it can be seen between the ducks.

tension, which is not very strong, can no longer hold each drop together. The big drops break up into smaller drops. There is a maximum size for raindrops. The largest raindrops are just over 1/8 inch (4 millimeters) across.

A little drop breaks off the top of the spike...   and keeps rising.   The spike falls back into the water...   leaving the little drop to fall in later.

Looking Through Raindrops

A raindrop hanging on a twig is a type of **lens**. Look into it to see a tiny, upside-down image of what lies beyond it. A lens **focuses** light so that objects look bigger or smaller, depending on how far the lens is from the object.

When the Sun shines, raindrops that are hanging from bushes and trees also act as **prisms** that glitter with the colors of the rainbow. One drop sparkles a brilliant blue, another sparkles red, another green. If you move your head slightly, each drop will change color. This happens because sunlight is a mixture of all the colors of the rainbow. Each raindrop separates the colors at slightly different angles. The color that is visible depends on the angle between the Sun, the raindrop, and you. When you move your head, the angle is slightly changed.

Top: A raindrop caught in a nasturtium leaf makes an image of the veins in the leaf.

Opposite: A raindrop hangs from a pansy bud. An upside-down image of the two pansies beyond is visible in the raindrop. The raindrop acts as a lens.

Left: Raindrops on leaves and twigs glitter with brilliant colors in the sunlight. Each drop is a tiny prism that splits light from the Sun into the colors of the rainbow.

Waterproofing

Rain is possible because air and water are able to mix. On the other hand, oil and water cannot mix. This allows many animals to use oil to waterproof themselves so that they stay dry.

Birds use oil to waterproof their feathers. Most birds have a special oil **gland** above the tail. When a bird **preens**, it takes oil from this gland with its beak and smears the oil over its feathers.

Right: With its beak, a barn owl takes oil from its oil gland and smears the oil over its feathers to waterproof them. This owl's oil gland, a little pink spike, is visible just above the tail.

This thin film of oil makes raindrops roll off the bird's feathers rather than soaking them.

Furry mammals also stay dry in the rain. Oil glands located in their skin keep their coats sleek and waterproof.

Certain parts of plants also need to stay dry. If water gets into plant leaves and stems, it **dilutes** the **sap** and damages the **cells.** If this happens, the plant cannot grow properly. To keep water out of these areas, many plants have a thin layer of natural wax on them. The wax also helps keep the plant from losing water from its sap during dry weather.

Above: These rats have just had a swim, and water drips off their fur. Oil from glands in their skin keeps the rats' fur waterproof.

Below: In dry weather, a wood sorrel leaf is wide open.

But heavy rain can damage the delicate leaf. So when the rain starts, the leaf begins to close.

Heavy rain cannot harm the leaf when it is closed.

Rain to Drink

Top: During the rainy season, the banjo frog comes out of hiding.

All plants and animals have water in their bodies. Plants draw water from the ground through their roots. Many types of animals drink water. Some soft-bodied animals, such as frogs, snails, and slugs, do not need to drink water because they are able to take in water through their skin. Because their skin is **permeable** to water, these animals can only be active in damp conditions. If they come out when the weather is dry, their bodies will dry up.

Very few living things survive for long without water. Yet some plant seeds can remain dry for years and begin to grow as soon as they are dampened by rain. The eggs of fairy shrimp may blow around in the desert dust for up to twenty years and still be able to hatch if they happen to land in a pool of water.

Above: The garden snail takes water in through its skin, but water can also evaporate through skin. So, in dry weather, the snail seals itself into its shell to keep its body from drying out.

Below: Rain is often scarce on the plains of Africa, but animals can smell rain from miles (kilometers) away. These zebras, wildebeests, and springboks may have traveled a long way to drink from the rain that was left after a heavy overnight shower.

21

Water at Work

Rain may fall as a fine drizzle or as a steady downpour of big drops. It may be a shower that lasts only a few minutes, or it may go on for days. Wherever rain falls over land, it drains mainly into streams and rivers that take it to the sea.

Much of our planet's surface has been shaped by rainwater. Rivers, swollen by rain, carve deep gorges down mountainsides. On level ground, rivers spread out and **meander**, making wide valleys. Rivers also carry **silt** to the sea.

Rain cleans the atmosphere, washing out dust and smoke particles from the air. Occasionally, rain washes so much **pollution** out of the atmosphere that it falls back to Earth as **acid rain**. Acid rain damages and kills plants and animals.

Top: Rivers shape giant rocks and small pebbles by grinding one against the other. Pieces get chipped off, and surfaces are smoothed. In this illustration, water-worn wood, seed heads, and the bleached body of a dung beetle have been left behind by floodwaters.

Opposite: When heavy rains fall, rivers like the Tana in northern Kenya in Africa become brown and swirling. Huge quantities of silt and soil are carried away in the water and dumped farther downstream or even into the sea.

Rainy Seasons

Top: A dwarf cactus survives desert **droughts** by storing water in its round stem.

Rain is so important to plants and animals that many of their lives are controlled by the rainy seasons. In many warm parts of the world, the rainy season is like spring in other areas. Seeds grow, plants send out new leaves, birds sing and build nests, and insects emerge by the millions.

Not all areas of the world have a regular rainy season. Rain in the desert can be very irregular, with some places having no rain at all for years. Only a few very **specialized** plants and animals can survive there.

Right: A cold rain falls on a baby warthog. The rain is cold because it has fallen from clouds that are high in the sky where the temperature is near freezing.

Other parts of the world receive rain during all the seasons. In a tropical rain forest, it may rain almost every day — usually in the afternoon. There, hundreds of different types of huge trees grow together, draped with vines, ferns, mosses, and orchids. Numerous types of colorful birds and insects live among the treetops.

Above: At the start of the rainy season in southern Africa, rain falls in heavy showers only in certain areas, leaving much of the land dry. Giraffes and impalas travel to places where rain has fallen.

The Water Cycle

Below: As you watch raindrops trickle down a window, imagine that the same water, perhaps only one or two days earlier, was part of the sea. Fish and even whales could have been swimming in it!

There is a saying that what goes up must come down — and this is true of water. Water goes up as vapor and comes down as rain. Water goes around — from the sea into the air, into rain, to rivers, and back to the sea again. That is why the process of rainmaking is called the **water cycle**.

The water cycle keeps water moving over the surface of Earth. The Sun provides the energy needed for this work. Its heat evaporates water from the surface of the sea, turning saltwater into freshwater vapor. The Sun creates wind that carries the vapor over land. Clouds form, and rain falls. Every living thing on land needs rain. Plants cannot live without water. Animals would not have plants to eat without water. Every animal needs water — for their bodies or for their watery homes.

Opposite: A shaft of sunlight shines through a gap in the storm clouds. Falling raindrops separate it into the colors of the rainbow. To make a rainbow, the Sun and rain are needed at the same time. The Sun and rain are parts of the water cycle.

Activities:

Measuring and Making Rain

Rain may fall heavily for a few minutes, or it may come down in a steady drizzle for hours. The effect rain has on the land depends mainly on the amount of water that falls. Rainfall is measured in inches or millimeters with a rain gauge.

To make a rain gauge, you will need a plastic funnel, a clear plastic glass almost the same size as the top of the funnel, and a ruler.

Put the funnel into the glass so that it fits tightly (see below). The funnel will keep rain from splashing out of the glass or from evaporating. Place your rain gauge in a quiet spot away from any trees or buildings.

Each day at the same time, measure the depth of water in the glass, and write the measurement down. Then empty the glass.

On some days, there may be no rain at all. On others, it may rain quite a bit. Keep a long-term diary of rainfall measurements to see how the rainfall amount varies from one time of year to another. Throughout the world, people record rainfall amounts. From their records, rainfall maps are drawn, showing the areas of heavy, medium, and light rainfall.

The Water Cycle in a Bowl

If you want to see how fresh rain comes from the salty ocean, make a water cycle in a bowl! You need a glass bowl, a short drinking glass, and some plastic cling wrap. You also need a spoonful of salt and some green or blue food dye.

Put some hot — but not boiling — water into the bowl. There should be enough water so that the drinking glass will sit on the bottom of the bowl without floating. Stir in the salt plus a few drops of dye. Place the drinking glass right side up in the middle of the bowl. Now you have an island surrounded by the sea (see photo).

Stretch a layer of plastic wrap over the top of the bowl and smooth it down the sides so that the bowl is airtight. Squeeze a little air out by pressing down gently on the plastic wrap while raising a corner of it. This should produce an indentation in the wrap. Then wait and watch.

Water is evaporating from the warm "sea." When the vapor reaches the cool plastic wrap, it condenses on it in tiny drops. The drops slowly get bigger until they join

branch in so that its cut end reaches the bottom and the tape is secure on the top of the bottle.

Put the bottle in a light and airy spot outside. Each day, check to see how much water the leaves have drawn up and turned into vapor.

If you know how much water disappears from the bottle each day and how many leaves there are on your branch, you can calculate how much water per day each leaf is "breathing" out. Multiply this by a million — because there may be a million leaves on a large tree – and you will know how much water a large tree puts into the atmosphere.

together to form a drip that falls into the drinking glass. "Rain" is falling on the "island!" When the rain is dripping steadily, pour some ice-cold water on top of the wrap, and watch what happens. Raindrops should form more quickly. You can see that the rain is not colored with the dye. Also, by tasting the water in the drinking glass, you will find that the rain is made of fresh water, not saltwater.

Leaves Make Rain

Water vapor in the air does not come just from the sea. It also comes from the land — particularly land that is covered by leafy plants.

To see how leaves put water vapor into the air, you need a clear glass bottle. Have an adult help you cut a small, leafy branch from a large tree. Place some tape around one end of the branch so that when its cut end is pushed to the bottom of the bottle, the tape tightly fits the neck of the bottle. Fill the bottle with water, and push the

Below: A parrot licks moisture from large forest leaves, while the leaves breathe moisture into the air.

Glossary

acid rain: rain that contains high levels of pollution.

atmosphere: the layer of gases that surrounds a planet.

atoms: tiny building blocks from which all substances are made.

cells: the microscopic building blocks found in living things.

chemical formula: letter and number symbols that identify certain chemicals.

condenses: changes from a vapor to a liquid.

dilute: the process of adding water to something.

drought: a long period of time when there is no rain.

focuses: adjusts to produce a clear image.

foliage: vegetation; plants.

gland: a part of an animal's body that produces a special substance to regulate a function in the animal's body.

hydrogen: a gas that is very light and burns easily.

lens: a piece of clear material with curved sides that focuses light.

meander: to wander from side to side.

molecule: a combination of two or more atoms.

oxygen: a gas that is needed by plants and animals to live.

permeable: a quality that allows water to soak through.

pollution: harmful substances in the air, soil, and water.

preens: smooths and arranges neatly, as when a bird smooths and arranges its feathers.

prism: a block of glass, a raindrop, or another transparent material that separates light into the colors of the rainbow.

rain forest: a forest that grows where there is an abundance of rain — usually in tropical regions of the world.

relative humidity: a measure of the amount of water vapor in the air.

sap: the liquid inside plants that nourishes them.

saturated: the condition in which an object is carrying the maximum possible amount of water or water vapor.

silt: fine particles in or deposited from water.

specialized: designed and suited for a specific purpose.

surface tension: the layer of molecules at the surface of a liquid that acts like a stretched elastic covering.

suspended: hanging in air or water.

tropical: coming from the warm regions around the Equator.

vapor: a gas formed from a liquid.

water cycle: the journey of water from the sea to land back to the sea.

Plants and Animals

The common names of plants and animals vary from language to language. But plants and animals also have scientific names, based on Greek or Latin words, that are the same the world over. Each plant and animal has two scientific names. The first name is called the genus. It starts with a capital letter. The second name is the species name. It starts with a small letter.

banjo frog (*Limnodynastes dorsalis*) — Australia 20

barn owl (*Tyto alba*) — worldwide 18

brown rat (*Rattus norvegicus*) — worldwide 19

common zebra (*Equus burchelli*) — southern and eastern Africa 20-21

cotoneaster (*Cotoneaster bullatus*) — China 14

dwarf cactus (*Rebutia calliantha*) — South America 24

garden snail (*Helix aspersa*) — Europe 21

giraffe (*Giraffa camelopardalis*) — Africa 25

greater flamingo (*Phoenicopterus ruber*) — Africa, India, southern Europe 10

green iguana (*Iguana iguana*) — South America 7

harlequin bug (*Philia senator*) — Australia 6

impala (*Aepyceros melampus*) — Africa 25

ladies mantle (*Alchemilla mollis*) — Romania, Asia Minor 30

mallard (*Anas platyrhynchos*) — Europe, North America 15, 18

nasturtium (*Tropaeolum majus*) — South America; worldwide 17

pansy (*Viola tricolor*) — cultivated worldwide 16

silver birch (*Betula pendula*) — Europe, Asia Minor, United States 8

springbok (*Antidorcas marsupialis*) — southern Africa 20-21

warthog (*Phacochoerus aethiopicus*) — Africa 24

white-lipped snail (*Cepaea hortensis*) — western Europe 3, 5

wildebeest or **brindled gnu** (*Connochaetes taurinus*) — southern and eastern Africa 20-21

wood sorrel (*Oxalis acetocella*) — Europe, Japan 19

yellow-fronted Amazon parrot (*Amazona ochrocephala*) — South America 29

Books to Read

Ask Isaac Asimov (series). *What Causes Acid Rain?* Isaac Asimov (Gareth Stevens)

The Audubon Society Guide to North American Weather. David McWilliams Ludlum (Knopf)

Rain. Jacqueline Dineen (Bookwright)

Rain. Andres Llamas Ruiz (Sterling)

Rain and Hail. Franklyn Mansfield Branley (Crowell)

Rain, Snow, and Ice. Ann Merk (Rourke)

Thunderstorm. Nathaniel Tripp (Dial Books for Young Readers)

Wonderworks of Nature (series). *Storms: Nature's Fury.* (Gareth Stevens)

Videos and Web Sites

Videos

Living Things in a Drop of Water. (Encyclopædia Britannica Educational Corporation)

Our Environment: Water: What Happens To It. (Agency for Instructional Technology)

Our Watery World. (Agency for Instructional Technology)

Web Sites

www.ns.doe.ca/aeb/ssd/acid/acidfaq.html
users.aimnet.com/~fundnat/RainForest Facts.html
www.eastnc.coastalnet.com/weather/nwsm hx/tstms.htm
www.ns.doe.ca//udo/here.html

Index